Sustainable Water and Waste Water Systems

RYAN BRAUTOVICH

Copyright © 2014 Ryan Brautovich

All rights reserved. No part of this book may be reproduced or transmitted in any form or by any means, electronic or mechanical, including photocopying, recording, or by any information storage and retrieval system, without the written permission of the Publisher.

Printed in the United States of America

November 2014

ISBN: 978-0-9864404-8-9

"We forget that the water cycle and the life cycle are one."

~ Jacques Cousteau

The Construction H.E.L.P. Foundation's Home Construction Audit program makes it easy and painless – through the use of our Home Building System – to understand how to build a home, how to manage your contractor, and how to protect yourself from being taken advantage of and scammed. We demystify the process and remove all of the contractor jargon to give you the building process in easy-to-understand, plain English. The Construction H.E.L.P. Foundation's founder and building expert Ryan Brautovich's exclusive 4-step home building system will ensure you are on the right track – and on budget – every step of the way. For more information about the Construction H.E.L.P Foundation, the Home Construction Audit Program, or any of the educational products, homeowner services, or construction seminars available in your area, please visit **www.HomeConstructionAudit.com**, or **www.ConHelp4U.org**.

Table of Contents

- Introduction 1
  - Codes and Requirements 2
  - Water supply 2
  - Sustainability considerations 3
- Water Supply Main Or Rainwater? 3
  - Code requirements For Rainwater 4
  - Designing and installing a rainwater system 5
- Harvesting Rainwater 5
- Roofing materials 6
  - Roof paint 7
- Gutters and downpipes 7
  - Leaf screens and guards 8
  - Leaf screens 8
- Diverters 8
- Storing rainwater 8
- Tank types and sizes 9
  - Tank materials 9
  - Filtration systems 9
- Backflow prevention 10
- Maintenance 10
- System Layout and Pipe work 10
- Water pressure 11
  - Water flow rate 12
- System Layout 13
- Backflow 14
- Water supply main connection 14
- Pipe materials and specifications 14

Pipe materials 14
    What to consider 15
    Copper 15
    Unplasticised polyvinylchloride (PVC-U or uPVC) 16
    Polyethylene (PE or HDPE) 17
    Polypropylene (PP) 17
    Cross-linked polyethylene (PEX) 18
Pipe Jointing Systems 19
Pipe work Installation 20
    Responsibilities 20
    General Installation Requirements 20
    Where To Lay Pipework 21
Access For Maintenance And Replacement 21
Preventing Electric Shock 21
Pipe Insulation 22
Noise and Air Locks in Pipework 22
    Water Hammer 22
    Air Locks in Water Supply Pipework 23
Backflow Prevention 24
Causes of Backflow 24
    Use an Air Gap to Prevent Backflow 25
    Backflow Prevention Devices 26
    Vacuum Breaker 26
    Double Non-Return Valve Assembly 27
    Reduced Pressure Zone Device 28
Installation Requirements 28
    Testing 28
    Valves and controls 29

- Hot Water Supply 30
  - Controlling temperature 30
  - Controlling Pressure 31
  - Water Heating Options 31
  - Other Considerations 31
- Storage Cylinders 32
  - Cylinder Size 32
  - Cylinder Location 33
  - Cylinder Insulation 33
- Controlling Bacteria and Tempering Heated Water 34
  - Tempering Heated Water 34
  - Specific Requirements For Gas Storage Water Heaters 35
  - Controlling Pressure in Storage Cylinders 35
  - Low Pressure, Open-Vented, Header Tank System 35
  - Low Pressure, Pressure-Reducing Valve System 36
  - District Main Pressure, Unvented Systems 36
- Hot Water Pipes 37
  - Materials 37
- Toilets 37
- Other Fixtures 38
- Appliances 39
- Wastewater 39
- Building Design Considerations 39
  - Common Plumbing Coordination And Installation Issues 39
- Sanitary Plumbing Systems 40
  - Types of fixtures 40
- Water Traps with Single Discharge Outlet 41
  - Water Traps with Multiple Discharge Outlets 41

- Access points 42
- Venting Discharge Pipes 42
- Toilets – discharge pipes and venting 42
- Air Admittance Valves 42
- Floor Wastes 42
- Pipe Materials 43

Drainage Systems 43
- Traps 43
- Ventilation of drains 43
- Pipe Sizing and Gradient 44
- Materials For Drains 44
- Testing Of Drains 44
- Access for Maintenance 45

On-Site Wastewater Treatment 45

Blackwater and Greywater 46
- On-Site Wastewater Treatment Options 46

Designing an On-Site Wastewater Treatment System 46
- Site considerations 47
- System Capacity 47

Design Checklist 48
- Stage 1: Feasibility Study 48
- Stage 2: Evaluation And Investigation 48
- Stage 3: Selection System 49
- Stage 4: Design Distribution System 49
- Stage 5: Approvals 49
- Stage 6: Installation 49
- Stage 7: Operation And Maintenance 50

Septic Tanks 50

How Septic Tanks Work 50
Construction and installation 51
Aerated and Advanced Wastewater Treatment Systems 51
How an AWTS Works 52
Land-Application Disposal System 53
Gravity Soakage Trenches/Beds (Septic Tank System
  Only) 53
Low Pressure Effluent Distribution (LPED)/Dose
  Loading (Septic Tank System Only) 53
Drip-Line Irrigation Systems 54
Evapo-Transpiration Systems (ETS) 54
Sand mound systems 55
Site Investigation 55
Maintenance and Problems 56
Maintaining Septic Tank Systems 56
Maintaining Aerated Water/Advanced Sewage
  Treatment System 56
Maintaining Land-Application/Disposal Systems 57
Avoiding Problems 57
Signs of Trouble 58
Solving problems 58
Greywater Recycling 58
Safety Considerations 59
Greywater Systems 59
How a Greywater System Works 59
Treatment Of Greywater 61
Designing A Greywater System 61
Proprietary Greywater Systems 62

 Irrigating With Greywater 63

 Local Authority Restrictions 63

 Safety and Health 63

 Distribution Systems 64

Composting Toilets 64

 How Composting Toilets Work 64

 Requirements of Composting Toilet System 65

 Deciding To Install a Composting Toilet 65

Types of Composting Toilet 66

 Batch-Type Units 66

 Continuous Systems 67

Stormwater Control and Landscaping 67

Controlling Stormwater Runoff 68

 Rainwater Storage 68

 Permeable Paving 69

 Swales 69

Green Roofs 70

 Reducing Garden Water Use 71

Conclusion 71

BIBLIOGRAPHY AND REFERENCES 72

Introduction

It is very easy to take water for granted when all you need to do is turn the faucet to get access to it. But surely you too have wondered where the water truly comes from while brushing your teeth in the morning. Is it really clean and safe? What is the source of the water? Has it been treated? If yes, then are the chemicals safe enough or the pipelines clean enough to make the water usable for consumption? The answer to these questions lies on one basic factor; your area of residence. Almost seventy percent of the water making its way in residential and commercial establishments is 66 percent treated surface water while only 34 percent is treated ground water.

Another factor that determines the quality of water running through your taps is the water supply system. If the water supply system is well constructed and well maintained, then the quality of water too would be safe and usable, if not then the purest of water sources cannot guarantee clean water at your doorstep., unless of course you are getting water from a private system using a well and a pump.

Water is a necessity and a blessing that should not be taken lightly. By minimizing water use, you can make good material choices, reduce running costs, cut demand on community infrastructure and even reduce the harm to the environment.

Reducing water use benefits the environment by, for instance, reducing the need to draw more water from rivers and waterways, reducing demand for energy, and reducing the need to build new infrastructure for supply and disposal.

There are numerous factors that need to be considered when designing a water supply and drainage systems; the building users' health and safety being the most pivotal one.

Codes and Requirements

A system is not sustainable if it does not meet basic needs for drinkable water and for safe disposal of waste. But, subject to health and safety requirements, it is good practice to design systems that support efficient and sustainable use of water, energy and materials. The costs of wastewater are borne by the building owners and occupiers through set rates and other user charges.

For water supply, there are many things to consider, ranging from the type of water heating used to the layout of pipework to specifying appliances and fixtures. These decisions can have a significant impact on water and energy use over the life of a building.

For wastewater treatment, the key decision is whether to connect to the main sewerage network, treat wastewater on-site, or a combination of both. Either way, it is important that health and safety requirements are met. Storm water runoff places demand on infrastructure and can carry contaminants into waterways such as streams and the sea. By designing, building and renovating homes that use water efficiently, you can help keep these costs down.

Water supply

Design a safe water supply system that meets building users' requirements while also making efficient use of water, energy and materials.

All habitable buildings are required to have a water supply that is potable (drinkable). That water supply must be protected from contamination, and must not contaminate the water supply system or source. As part of the water supply system, the building must also provide appropriate facilities for personal hygiene, and washing utensils. Hot water should also be provided that is safe and will not cause scalding. All water must discharge to a wastewater system to safeguard people from illness and protect them from odors and waste matter.

Sustainability considerations

While the first consideration in designing a water supply system is the building occupiers' health and safety, it is good practice to design systems that support efficient and sustainable use of water, energy and materials.

There are sustainability considerations in most aspects of water supply, including decisions about the sources of water, location and layout of pipework and storage, materials used, heating methods, and appliances and fixtures specified.

Water use can be minimized by controlling water pressure, providing an efficient hot water system, controlling the water used for toilet flushing, specifying efficient appliances, minimizing outdoor water use and reducing the water flow from the outlets.

Water Supply Main Or Rainwater?

Most buildings use a connection to district water supply mains. However, rainwater is an option to partly or fully replace water main systems. Water main systems offer reliable supply that meets required standards for potable (drinkable) water.

For most buildings where water mains are available, using the main supply will be the most effective way to meet users' water needs. Other features – such as specifying efficient fixtures and appliances, and designing an efficient water heating system – can be used to reduce the building's environmental impact.

But rainwater can be used to provide water supply when there are no main connections available. Even if there is a main connection available, rainwater can be used either to meet all of a building's demand for water (depending on users' water needs) or to meet some water supply needs such as providing water for gardening, flushing the toilet, and/or bathing and clothes washing.

Using rainwater can:

- reduce costs for users on a metered water supply
- provide an emergency supply
- reduce demand on main water supplies
- reduce demand on storm water disposal systems
- provide an independent supply for watering the garden in times of drought
- reduce the rate of storm water runoff

In the United States, there are over 100,000 systems that the population depends on for roof-collected rainwater for drinking water.

Code requirements For Rainwater

Any rainwater system must meet relevant Building Code requirements. This includes a requirement for adequate potable (drinkable) water to be provided for consumption, oral hygiene, utensil washing and food preparation. This potable water supply must be protected from contamination, and must not contaminate the water supply system or source. Potable water must meet the relevant standard for drinking water.

The Building Code also requires adequate water supply to any sanitary fixture (such as toilets, baths, showers, sinks and so on). The sanitary plumbing system must be set up to minimize any risk of illness or injury.

Any rainwater system that is connected to a main water supply must be designed to minimize the risk of contamination of the main water supply by including an air gap or backflow prevention device. The system must also be designed to minimize the risk of contamination to rainwater intended for household use.

Designing and installing a rainwater system

The design and installation of any rainwater system will depend on its purpose. If you require a small rainwater barrel to water your garden, or a larger storage system for sanitary uses or potable water supply. In general, rainwater will be harvested from the roof and stored in a tank until use. The system needs to be designed, installed and used to minimize the risk of contamination.

Irrespective of whether a connection to the water main or rainwater is used, the designer will need to consider how to manage water pressure, and pipework layout and specifications, as well as how to achieve efficient water heating.

Harvesting Rainwater

It's crucial to ensure that the roof, guttering, pipes and other features do not contaminate the water before it goes into the storage tank. Water from roofs can be contaminated by decaying vegetable matter such as leaves, petals and pollen; fecal matter from birds, possums and rats; and dead insects, birds and animals. Other sources of contamination include particulates from solid fuel flues; pollution; chemical spray drift; and harmful materials – such as lead – in the roof, roof paint, gutters, downpipes etc.

The Plumbing Code requires, among other things, that water supplies must be protected from contamination. Check your local Plumbing Code for water supply main, or rainwater requirements for rainwater collection.

Following are a list of ways you can reduce the risk of water contamination:

- specify roof, spouting and pipework materials that will not contaminate the water supply
- specify leaf guards over the gutters and leaf screens on downpipes

- specify a first flush diverter to prevent the first 5 gallons of water, which may be heavily contaminated, from entering the storage tank
- do not collect rainwater from a roof that has overhanging trees or a TV aerial – this reduces the risk of leaves and bird droppings getting into the harvested rainwater
- ensure that the flue from a solid fuel burner is located so that soot and other discharges are carried clear of the rainwater collection area
- advise owners that gutters must be kept clean

If rainwater is being harvested for human consumption, roofing, spouts, downpipes, and pipework materials must comply with local regulations.

Roofing materials

Roofs suitable for water collection for human consumption include:

- unpainted zinc/aluminum-coated or galvanized steel
- factory-coated or painted zinc/aluminum alloy-coated or galvanized steel
- zinc
- stainless steel
- aluminum
- concrete or terracotta (clay) tiles
- copper
- PVC (without lead stabilizers) or fiberglass sheet
- untreated timber shingles (usually imported western red cedar)
- butyl rubber.

When purchasing roof materials, it is imperative to keep into consideration the health and safety. However, you should also take account of sustainability considerations like embodied energy. Ideally, you should leave a new roof for one good period of rainfall before connecting the downpipes to the storage system.

You should not use collected water for drinking if it has come into contact with:

- uncoated lead flashings (lead flashings on existing roofs should be coated with a suitable paint; coated lead is available for new roofs)
- treated timber where chemicals leaching out might contaminate the water
- bitumen-based roofing
- asbestos (although no longer used in building, existing asbestos roofs should not be used for collection of rainwater).

Roof paint

Specify only paint that the manufacturer recommends as suitable for collection of rainwater for drinking. Do not use collected water for drinking from roofs coated with lead-based paints or bitumen-based paints. Also do not use collected water from roofs coated with acrylic paint until it has been washed by a good rainfall.

Gutters and downspouts

Materials suitable for use for gutters and downspouts where potable water is being collected include extruded PVC, factory-coated zinc/aluminum alloy-coated steel, galvanized steel, copper (which may cause staining if water has a low pH), aluminum, and polyethylene/polypropylene.

Again, while the first consideration is health and safety, you may also want to take account of sustainability considerations such as embodied energy. See material use for more.

Leaf screens and guards

Leaf screens and guards help keep plant matter and other debris out of the rainwater storage tank.

Leaf screens

Leaf screens located on each downpipe keep larger debris out of the rainwater tank.

Diverters

These devices divert the first rain away from the water collection tank, washing dust, leaves and other debris off the roof before water is collected.

One form of diverter has a float that rises as the rainwater flows in. When the floater reaches the top of the diverter pipe, it seals it off allowing the rainwater to flow into the tank. Generally, the more water that is diverted, the better the quality of the collected water.

Storing rainwater

The tank for storing rain water should be suitable for the purpose it is built for. If the water is to be used for drinking, then it should be able to minimize the risk of contamination. If it is to be used for gardening purposes alone, then it should at least store water properly without leakage or damage to the tank.

Once rainwater has been harvested, it must be stored in a tank for use. Tanks are available in different sizes – from small barrels for gardening water to tanks that are large enough to cover all of a building's water needs.

To reduce the risk of contamination:

- site the tank, if possible, so that it is shaded from the sun, particularly during the hottest time of the day

- specify tightly fitting covers for all tank inspection ports, insect screens on all vents and openings, and an overflow that can siphon out fine sludge
- specify an intake near the water surface or a floating intake, with optional filter to draw the water from the best area in the tank
- specify a vacuum overflow that clears debris from the bottom of the tank.

Tank types and sizes

There are several types and sizes of rainwater tanks, designed to meet different needs. The tank selected should provide adequate supply for the purpose, whether that is irrigation, toilet flushing or all household use.

Tank materials

Suitable materials for water storage tanks include galvanized steel, fiberglass and plastics including polypropylene, and concrete. Tank materials should not transmit light, as light will encourage organic growth. When specifying materials, while the first consideration is health and safety, you may also want to take account of sustainability considerations such as embodied energy.

Filtration systems

Filtration systems may be:

- point-of-use (attached to a tap or plumbed in with a dedicated faucet), or
- point-of-entry (centrally installed system to treat all water).

Types of filters include:

- mesh filters of various sizes to remove different types of particles
- carbon filters

- reverse osmosis filter
- UV sterilizers to kill bacteria

Ideally, a filtration system should include a number of different types of filters.

Backflow prevention

As was noted in water supply mains or rainwater, any rainwater system that is connected to a main water supply must be designed to minimize the risk of contamination of that specific main water supply.

Acceptable solutions provide that there must be no likelihood of cross-connection between a private water supply system (such as a rainwater system) and main water supply. This can be achieved by using an air gap or a backflow prevention device such as a double non-return valve.

Maintenance

Rainwater systems should be maintained following a program that includes:

- **Three Monthly Inspection** – clear gutters, leaf guards and first flush diverters; clean and replace filters as necessary; trim back trees overhanging the roof
- **Annual Inspection** – clean tank to remove accumulated sediment; sludge may be removed by siphoning or pumping without requiring the tank to be emptied.

The manufacturer's recommendations for tank and filter maintenance should be followed, and tank water should be treated to remove organic materials if necessary.

System Layout and Pipe work

The water supply system must be designed to achieve appropriate water pressure and flow, and to avoid contamination to potable water.

As well as avoiding contamination and achieving the right pressure and flow, the system must be suitable for the temperature of water carried. A well-designed and installed system will also be durable, minimize noise from water flow and from problems such as water hammer, and support efficient use of water.

All water supply systems use a combination of pipes (of different dimensions and materials), valves and outlets to deliver water to building users. Some water supply systems also use storage tanks and pumps. Designing a water supply system involves getting all of these elements right so that clean water is delivered to the user at the appropriate rate and temperature.

Water pressure

It is crucial to have the right water pressure to use the water efficiently. Too low water pressure can become inconvenient for the building users as the showers would have a poor water flow and the baths would take much longer to fill. At the same time, if the water pressure is too high, it can lead to wastage of water and high wear tear on the system.

Typically, new buildings in areas with water supply mains will have pressure systems. Existing buildings, and buildings that are not connected to water supply mains, may have low pressure systems or unequal pressure systems (with different pressures for hot and cold water supply).

Water main pressure systems require pressure limiting and pressure reducing valves to control water pressure and temperature. Typically, pressure limiting or pressure reducing valves will be used to control pressure in main into the house or where high pressure may lead to problems such as burst pipes.

Low pressure systems require few valves or controls. In low or unequal pressure systems, pressure can be increased to adequate levels by storing water in a header tank (typically in the ceiling space) so that gravity can be used to create water pressure.

Pressure can also be raised to adequate levels using a pressurizing pump, in which case it may be necessary to use pressure limiting and pressure reducing valves.

Water flow rate

The Building Code requires that sanitary fixtures and appliances have adequate water supply at an adequate flow rate.

As with water pressure, flow rates are crucial. A flow rate that is too high will result in water being wasted, whereas a flow rate that is too low will mean that sanitary fixtures and appliances don't work properly.

Flow rate is affected by:

- Water pressure
- Pipe diameters – The smaller the internal diameter of the pipe, the lower the pressure and flow rate.
- Pipe lengths – longer pipes will result in a lower flow rate
- Number of bends and fittings – the more bends in a length of pipework and the more fittings, the lower the flow rate
- Water temperature – higher temperatures will tend to raise pressure and flow rates

A flow regulator can be used to maintain a constant flow, independent of water pressure. As an instance, if someone is in the shower and the kitchen tap is turned on full, the temperature and flow are likely to remain more stable if a flow regulator is used.

Limiting the flow for a tap or appliance to a reasonable rate helps balance the available pressure throughout the system. Regulating flow allows a simpler design and minimum pipe sizes as peak flow rates can be specified accurately and can also reduce noise, splashing taps, and water hammer.

Manufacturers' recommendations must be referred to for pressure and flow information when selecting tempering valves and outlets (taps, mixers and shower heads).

Flow rate can also be controlled by specifying low-flow outlets.

System Layout

In the design process, the layout of the plumbing system will largely follow room layout. Nonetheless, there are many things to consider which relate to Code compliance, building users' comfort, and sustainability.

When planning a water supply layout, the following must be considered:

- **Pipe Runs And Lengths** – Keep pipe runs as short as possible. Pass pipes close to fixtures to minimize the number of branches and unnecessary elbows, tees and joints. Having longer pipe runs and more fixtures will reduce flow rate, increase heat losses, and increase use of materials.

- **Point Of Entry Into The Building** – This should be into a utility space such as garage/laundry and should include an accessible isolating valve, line strainer and pressure limiting valve (if required).

- **Water Heating System** – Locate centrally to reduce the length of pipe runs to fixtures because longer pipe runs require more water to be drawn off before hot water is discharged. Install a separate point-of-use water heater for fixtures that are more than 10 m from the main water heater.

- **Noise Prevention** – Avoid running pipes over or near bedrooms and living areas.

Backflow

Backflow is the unplanned reversal of flow of water (or water and contaminants) into the water supply system. The system must be designed and used to prevent contamination from backflow. See preventing backflow for more.

Water supply main connection

Where the water source is a main supply, the network utility operator is responsible for the water supplied to the property boundary. The property owner is then responsible for providing the pipework to bring the water into the building.

An isolating valve must be fitted at the point of connection to allow for maintenance and repair of the building's water supply system if required.

Pipe materials and specifications

The pipes used in a building must not contaminate potable water supply, and must be suitable for the water pressure, flow rate and temperature of water they will be carrying. This will be influenced by the materials used and also by other factors such as the wall thickness.

Other considerations are durability, ease of installation, cost, and sustainability. Common materials include copper, unplasticised polyvinylchloride (uPVC or PVC-U), polyethylene (PE), polypropylene (PP-3 or PP Type 3), and cross-linked polyethylene (PEX).

Pipe materials

Pipes must not contaminate water, and must be suitable for the pressure, flow and temperature of the water they are carrying. All of these are acceptable materials for water supply pipes.

What to consider

Pipe materials and components must not contaminate potable water. They must also be:

- suitable for the expected temperatures and pressures
- compatible with the water supply, to minimize the potential for electrolytic corrosion
- suitable for the ground conditions (if used underground) to minimize the potential for corrosion of the exterior of the pipe
- suitable for the local climate (if used outdoors) such as freezing conditions or atmospheric salt or sulfur
- able to withstand UV effects (if used outdoors)

So, when selecting of materials for water supply pipes, consider water pressure, water temperature, compatibility with water supply, durability, support, ease of installation, and cost.

You should also take account of sustainability considerations such as embodied energy.

Copper

Copper has long been used for all types of domestic water services and distribution because it:

- is durable
- has good corrosion resistance
- is malleable and easy to bend
- is self-supporting
- has good flow characteristics
- requires few fittings
- can be recycled

Copper may be annealed (i.e. heated, then cooled slowly) which improves its properties, for instance making it less brittle and stronger.

Although copper in general has good corrosion resistance, this depends on the environment. Acidic conditions, either from the soil (if buried) or from the water, can cause corrosion, so local pH levels should be checked before using copper pipes.

The use of copper pipes for drinking water supply is somewhat controversial. While copper is an essential trace element, high doses of copper (above 25 mg/lb body weight) can be lethal. The concentration of copper in tap water may increase if low pH water remains in contact with copper pipes and fittings. Households with very low pH water should follow these recommendations:

- Flush the system by running the water for 2–3 minutes or until the water becomes as cold as it will get, if a tap has not been used for approximately 6 hours or longer
- Flush each tap individually before using the water for drinking or cooking
- Do not use water from hot water taps for cooking or drinking, as hot water dissolves copper more readily than cold water does
- The flushed water can be used for washing or for watering plants.

Unplasticised polyvinylchloride (PVC-U or uPVC)

PVC-U is the most widely used plastic piping for water supply pipes and drainage pipes. It can be used in internal, exposed outdoor and underground situations as it:

- is durable
- is inexpensive
- has good flow characteristics

- has chemical resistance
- can withstand UV
- is easy to handle

It is suitable for cold water services and can be used for limited hot water services as it has a maximum temperature use of 140°F. Chlorinated polyvinyl chloride (CPVC or PVC) is produced by the chlorination of PVC. The result is a PVC pipe with a service temperature of 190°F that may be used for hot water supply pipes.

Polyethylene (PE or HDPE)

High density polyethylene (often called alkathene or polythene) has been used since the early 1960s. It is suitable for both potable water and wastewater services but it can only be used for cold water supply and is not suitable for hot water. It is the most commonly used plastic pipe for supplying the main water to a dwelling. Polyethylene:

- is durable
- is corrosion resistant
- has good flow characteristics
- is lightweight and flexible
- is easy to install
- has a good bending radius
- is inexpensive
- requires few fittings

Polypropylene (PP)

There are three types of polypropylene:

- P-H has good mechanical properties and excellent chemical resistance for use as industrial and sewerage waste pipes systems.

- PP-R has good resistance to high internal pressure so it is suitable for domestic pressure water supply systems and both hot and cold water services.
- PP-B is suitable for buried sewerage and wastewater drainage as it has good impact strength, particularly at low temperatures, and excellent chemical resistance.

Polypropylene has become increasing popular since the late 90s due to its:

- chemical and corrosion resistant
- heat resistant
- lightweight
- easy to install
- frost resistant

When used outdoors, polypropylene should be protected from UV exposure.

Cross-linked polyethylene (PEX)

PEX tubing is made from a cross-linked, high density polyethylene polymer, which results in a stronger material that polyethylene. Following are a list of PEX properties:

- It is more durability under extreme temperatures and chemical attack.
- It has greater resistance to cold temperatures, cracking and brittleness on impact.
- It can be used for hot water supply and hydronic heating systems, as well as potable water supplies.
- It is very flexible.
- It can be installed easily.

- It can be used for indoor and even outdoors when buried.

PEX is not recommended for outdoor above ground use – Although PEX can withstand some UV exposure, it is not recommended for outdoor above ground use. Make sure you read the manufacturer's instructions when installing PEX pipes.

Pipe Jointing Systems

The type of pipe jointing system used depends on the pipe material.

Brazing is the most common method for joining copper pipe in the United States. Straight joins are made by soldering using a solder that comprises copper, phosphorus and 15% silver, to create a lapped capillary joint that is permanent and durable.

Manipulative mechanical jointing uses brass fittings to make copper pipe connections. A nut is placed over the end of the pipe and a swaging (crox) tool is inserted to expand the pipe, creating a rolled groove to secure the nut in position. The joined ends are made watertight using plumbers' hemp or thread tape. This joint is most commonly used for connecting pipes to valves and fixtures. It is prone to loosening over time and should therefore not be used in concealed or inaccessible locations.

Non-manipulative jointing also uses brass fittings, but instead of expanding the pipe with a swaging tool, a brass 'olive' is placed over the pipe and compressed between the nut and fitting to create a secure joint that can easily be separated later.

Crimp ring uses an external stainless steel or copper ring that is placed over the pipe then compressed with a hand tool. As long as the ring is correctly placed and aligned, the resulting joint is very robust.

Sliding sleeve uses a sleeve that is placed over the pipe end and then expanded to go over the serrated spigot. A special tool forces the sleeve over the pipe and spigot to create an effective joint. Sleeves can be removed with the application of heat and then reused. The disadvantage of this connection is that the tools to create it can be difficult to use in confined spaces.

Heat fusion welding is where the surface of the pipe and connection are melted together using a heating iron. As the two ends are overlapped and fused without the application of welding fillers, the result is effectively a continuous pipe.

Solvent cement welding also overlaps and fuses the pipes but uses a solvent to 'glue' the pipes together.

Pipe Installation

Pipes must be installed to meet requirements for durability, safety and thermal performance.

Responsibilities

- Before a project starts, determine the responsibilities for specific work of the plumber, the main contractor and the electrician. Discuss the location of pipe runs with the main contractor to minimize cutting or notching of framing.

General Installation Requirements

Pipe must:

- comply with the durability requirements of Building Code
- be compatible with the support
- be installed to allow for thermal movement
- be protected from freezing by insulation, or being buried below the level of freezing
- be protected from damage

- be wrapped in flexible material or sleeved when penetrating masonry or concrete.

Where To Lay Pipe

Pipe may be installed:

- in a roof space
- under a timber floor
- below a concrete slab

Where water pipes are buried, they must have at least:

- 12 inches below finished grade
- 6 inches below frost line

Where pipes are under concrete, they should have 50-year durability.

Access For Maintenance And Replacement

Pipes installed in locations that are difficult to access should:

- have a detailed durability statement for 50-year service from the manufacturer
- if passing through a structural element, be sleeved in a larger pipe that is open at both ends to allow the pipe to be inspected or replaced if necessary
- be installed in a chase or duct which will provide ready access and will not compromise structural elements.

Preventing Electric Shock

Electric shock from water pipes may occur in any building where:

- the water supply piping is metal and in contact with the ground
- an electrical supply is provided into the building

- building occupants are able to make contact with exposed pipes

To avoid the potential of electric shock:

- the pipework must be connected to an earth electrode using earth-bonding conductors
- metallic fixtures must be bonded to the pipework

Safety precautions must be followed when cutting metal pipes.

Pipe Insulation

The plumbing code requires pipework to be insulated for a number of reasons and scenarios including:

- to limit heat loss (hot water pipes)
- to prevent water in pipes from freezing (hot and cold water pipes)
- all water pipes should be insulated except where connected to a heat dissipation device

Noise and Air Locks in Pipework

The Following are a number of ways to prevent water hammers and air locks:

Water Hammer

Water hammer (or pressure surge) generally occurs in a high pressure system when the flow of water is suddenly stopped. A sudden fluctuation in flow velocity sets up shockwaves through the pipework, causing the pipe to vibrate making a 'hammering' sound. It mostly occurs in metal pipes, although it can occur in plastic pipes also.

Fast-acting taps such as lever taps with ceramic disc washers, solenoid valves such as those on washing machines, spring-closing valves and pumps are often a cause of water hammer. It is related directly to the water velocity

– the faster the water travels, the greater likelihood of water hammer.

It is better to prevent water hammer than trying to fix the problem once a building is complete.

To reduce the likelihood of water hammer:

- avoid direct contact of pipes with the structure
- clip pipes with rubber insulated clips or clip over the pipe insulation
- fix pipework rigidly to prevent movement
- provide relief bends or flexible sections of pipe to absorb shock
- fit grommets or cushioned packers where pipes pass through structural members
- size pipework to avoid excessive water velocity (below 10 ft/s)
- limit system pressure – recommended is 50 – 70 psi

Air Locks in Water Supply Pipework

If air enters a water supply system, it will accumulate at high points and can restrict the flow of water. If there is not enough pressure to push the air bubble through the pipe, the air lock will remain until the pipeline is manually purged.

Air may enter the system from:

- a cylinder vent
- the tank if it runs low
- water as it is heated

Low pressure pipes should be graded to allow air to exit from predetermined high points and to prevent air locks from occurring.

Backflow Prevention

Backflow is the unplanned reversal of flow of water (or water and contaminants) into the water supply system. The water supply system must be designed and used to prevent contamination from backflow.

For backflow to occur there must be a physical connection, or cross-connection, between the water supply and any delivered water or contaminant. A common situation is the end of a garden hose submerged in a bucket or other container of liquid. Backflow can also arise from appliances, pools, and water storage tanks such as header tanks and rainwater tanks.

Advise building owners to take simple precautions such as not submerging garden hoses or spray heads from showers and sinks, and always turning off the water supply at the tap when it is not being used. A number of other methods to minimize the risk of backflow have been discussed below.

Causes of Backflow

Backflow is caused by a difference in pressure and may occur due to:

- **Backsiphonage** – the supply pressure is less than the downstream pressure, allowing water to be pushed in the wrong direction
- **Backpressure** – for instance, insufficient relief of pressure in a vessel where water is heated.

Backflow can only occur where there is a connection or cross-connection. Cross-connections can occur in any situation where fixtures are connected directly to the main supply such as:

- irrigation systems
- dishwashers
- washing machines

- coffee machines
- swimming pools, spa pools or ornamental pools that are filled by hose
- water softeners
- pesticide and fertilizer attachments for hoses
- fridges and icemakers
- bidets
- retractable spray outlets to tubs and sink
- flexible shower hoses
- storage tanks

Use an Air Gap to Prevent Backflow

In most situations, an air gap is the most cost-effective and reliable form of backflow prevention. An air gap should be used to prevent backflow from rainwater tanks and other water supply tanks into the main-supplied water system. Air gaps should also be used to prevent backflow of contaminants from all appliances and fixtures that are connected to the water supply.

For swimming and spa pools, provide a dedicated water supply with an approved air gap.

If main supply is used to top up a private water supply, backflow can be prevented by using a floating weight that can be used to operate a valve, ensuring that the maximum water level always remains at least 1 inch below the main inlet, or as required by your local municipality. Alternatively, a double non-return valve can be used.

If a piped supply is used to top-up the rainwater storage tank, a simple commercially available floating switch will ensure that top-up water is added only when the level in the tank is low. A float valve should not be used because it will add piped water whenever there is any draw-off.

If there is any direct connection between main supply water and a rainwater collection system, then a means of preventing backflow, such as a double non-return valve, must be designed into the system.

Backflow Prevention Devices

If the system is a high pressure system and a pipe is directly connected to an appliance or sanitary fixture, it may not be possible to use an air gap. In this case, a backflow prevention device must be installed.

The appropriate device for a particular installation can depend on the:
- hazard level of any potential contaminant
- potential for cross-connection
- type of backflow expected
- physical limitation of the device and the environment

Generally, the higher the hazard, the higher the risk, so the safer the device must be.

Vacuum Breaker

A vacuum breaker contains a float disc and an air inlet port. Under normal water flow, the float disc closes off the air inlet port, but if the normal water flow is interrupted, the float drops, closing off the system against backflow and, at the same time, opening the air inlet port.

A variety of vacuum breakers are available, including the following:
1. An atmospheric vacuum breaker (AVB) is one of the simplest and least expensive backflow prevention devices. It can provide excellent protection against backsiphonage. It consists of a gravity plunger or float disc that is forced upwards when the supply pressure is turned on, thus sealing off the atmospheric vent overhead. As soon as the supply is interrupted or terminated, the float drops down and opens the downstream pipework to

atmosphere. There must be sufficient pressure to fully lift and seal the float on the vent, so it is not suitable for use on very low pressure systems.

2. A hose connection vacuum breaker (HCVB) is a specialized type of atmospheric vacuum breaker designed to attach directly to the hose tap. It has a spring-loaded check valve that seals against an atmospheric outlet when the water supply is turned on. When the supply is turned off, the device vents to atmosphere, thus protecting against backsiphonage conditions. It is non-testable and should not be used as protection against backpressure or be subject to continuous pressure (2 hours maximum is permitted), i.e. no control valves should be located downstream of the device.

3. A pressure vacuum breaker (PVB) evolved due to a need to have a vacuum breaker that that can be subject to constant pressure and is able to be tested in line. It is similar to the atmospheric vacuum breaker except that there is a spring to hold the disc float in the open position during normal operation. They have two isolating valves and two cocks for testing, one for each chamber. These devices can be used under constant pressure but do not protect against backpressure.

Double Non-Return Valve Assembly

Essentially, a double non-return valve assembly (also known as a double check valve assembly or DCVA) consists of two independently operated non-return valves within one body. One non-return valve simply acts as a back-up. Because there is a risk that both valves will fail at the same time, regular testing is imperative, and the device is limited to use in medium and low hazard situations. This valve will protect against backpressure and backsiphonage but is not fail-safe. Because of the spring pressures, there can be a significant reduction in pressure across this device.

Reduced Pressure Zone Device

This backflow protection device incorporates two independently-acting, spring-loaded check valves separated by a differential pressure relief valve. Pressure between the two valves is lower than the supply pressure during normal operation. If either check valve leaks, the pressure relief valve will open, discharging water out of the system.

This device provides the maximum protection of any valve and can be used in high hazard situations.

Installation Requirements

All backflow prevention devices require a building consent for installation and must be:

- installed as near as practicable to the potential point of contamination
- protected from physical and frost damage
- isolated from corrosive or toxic environments
- installed above surrounding ground level so that leakage from air ports and discharge ports is readily visible
- installed in a position and manner to be accessible for maintenance and testing
- fitted with a line strainer upstream to prevent particles in the pipework from rendering the device ineffective
- attached only after the pipework has been flushed
- installed without the application of heat

Testing

Backflow prevention devices may be testable or non-testable. Their use in a particular situation depends on the degree of hazard. Non-testable devices may only be used on low-hazard rated systems.

Testable devices must be tested on installation and at regular intervals to the standard set down by the local municipality. Non-testable devices should be checked every 2 years maximum.

Valves and controls

Valves and controls are required – particularly in main pressure systems – to protect water supplies from contamination and to achieve the desired water pressure, flow and temperature.

- **Isolating valves** – are manually operated valves to isolate one section of a system.
- **Pressure limiting valves** – limit the pressure within a pre-set range
- **Non-return valves** – prevent reverse flow within or from the system
- **Line strainers** – Filter particles of solid matter from the water to protect other valves further downstream
- **Pressure reducing valves** – reduce the pressure to a pre-set level
- **Expansion control valves (cold water)** – Release pressure in the cold water feed pipe caused by the expansion of water in the storage water heater cylinder during normal operation.
- **Pressure relief valves (PRV)** – release pressure in the storage cylinder if it rises above the pre-set limit
- **Temperature pressure relief valves (TPRV)** – operate above a pre-set temperature and pressure
- **Temperature limiting thermost**ats - control the temperature of the water

Hot Water Supply

Hot water supply must be adequate to meet building users' needs while also keeping them safe. It should also support efficient use of both energy and water.

In designing a water heating system, the key decisions will include the source of energy for water heating, whether to use a storage cylinder or continuous flow system, system layout, and system capacity (including delivery rate, recovery rate, actual and potential number of users, type and number of fixtures within a household).

The system must be designed to meet safety requirements, which largely concern controlling temperature and pressure to ensure there is minimal risk of scalding or of a storage cylinder exploding.

A well-designed system will also minimize energy and water use, for instance by using an efficient heating source, ensuring the pipe runs are relatively short, and by using efficient fixtures and appliances.

Controlling temperature

Water heated to more than 120°F can cause serious burns in less than a minute. This creates risk, particularly for children (who have sensitive skin) and the elderly (who have slower reaction times).

Most Plumbing Codes require that, in most buildings hot water delivered to sanitary fixtures such as basins, baths and showers should not exceed 120°F (lower temperatures are required for some buildings such as hospitals, schools, and care facilities).

If a storage cylinder is used, the water must be heated to over 140°F at least once a day to prevent the growth of bacteria. The water must then be tempered to reduce the temperature before it is delivered to outlets.

Continuous flow systems are not at risk as they do not store water that would allow growth to occur.

Controlling Pressure

In buildings with water that is connected to a water district supply, a system of valves and controls will be required to regulate water pressure and temperature. See controlling pressure in storage cylinders and valves and control for more detail.

Water Heating Options

Probably the key decisions to be made are which energy source to use, and whether to use a storage system or a continuous flow one. In continuous flow systems, water can be heated using electricity or gas, while in storage systems, the water can be heated using electricity, gas, solar energy or a wetback.

Each energy source has its advantages and disadvantages, as do storage and continuous supply systems. In general, solar, heat pump and wetback systems are more energy efficient than traditional electric and gas systems. Continuous flow systems can be efficiently used in some circumstances, such as to boost a solar system, or to feed an outlet that is some distance from the main hot water supply.

Other Considerations

As well as specifying an efficient energy source, water and energy efficiency can be enhanced by:

- designing the system to minimize pipe runs – for instance, by locating the storage cylinder close to the kitchen, laundry and bathroom
- specifying low-flow fixtures
- specifying appliances that use water and energy efficiently

- locating the hot water storage tank close to kitchen and bathrooms
- installing a continuous-flow hot water system, particularly for remote outlets, so that hot water does not need to be stored
- insulating the pipes to reduce heating costs (but this has less impact than reducing pipe lengths)

Install a secondary continuous-flow system to reduce water wastage, if a kitchen is remote from the hot water storage system.

Storage Cylinders

Hot water storage systems can be used with energy-efficient heating sources such as solar, air-to-water heat pumps or they can use gas or electricity as the primary energy source. A disadvantage is that they can run out of hot water.

Cylinder Size

The hot water storage cylinder must be large enough to provide for a household's peak hot water demand, however, it shouldn't be too large so as to heat more water than required. The appropriate size depends primarily on the number of people in the household. Typical hot water usage is in the order of 45–65gallons per day per person.

Significant standing losses occur from the cylinder and during the transfer to the point of use. For greater efficiency, hot water storage cylinders should be short and broad rather than tall and slim, as this reduces the surface area.

Taller cylinders may be better where heat is exchanged to and from other sources (such as a solar collector or wetback) to the cylinder. For instance, a wetback may draw cold water from the bottom of the cylinder and return it as hot water to the top of the cylinder. A taller cylinder will allow a greater temperature difference to be maintained and therefore improves the effectiveness of the wetback.

Cylinder Location

Locate the cylinder as close as possible to the outlets where most hot water is used in order to minimize heat loss through the hot water pipes.

The cylinder will lose more heat in a cold location (e.g. when located outside the insulated building envelope) than a warm one (such as in the middle of the house or in a well-insulated space). Locating the cylinder in a cupboard will help to retain heat. Modern gas storage cylinders are often designed for installation outside the building envelope.

Other design factors to consider include:

- hot water system pressure – low or district main pressure
- provision of sufficient space for the cylinder
- access to replace the cylinder
- plan spaces where hot water will be used to be in close proximity – if an isolated hot water outlet is required, or hot water demand will be low, a continuous flow water heater may be a better option
- placement in relation to solar collectors or solid fuel burners where these are used for water heating

Cylinder Insulation

Maximize energy efficiency by:

- wrapping the cylinder with additional insulation
- insulating the hot water pipework

Electric storage water heaters should be insulated with an insulation jacket that has at least an R-value of 24; this will reduce heat loss by 25-45%. Ideally insulation should be added if the cylinders are warm to the touch.

Controlling Bacteria and Tempering Heated Water

To prevent the growth of bacteria, stored water should be heated to not less than 140°F at least once a day. If the thermostat control is set above 140°F, hot water storage systems are not at risk of growth.

If the hot water storage system is partially heated by solar power or heat exchange system (wetbacks or solar heat transfer system), the temperature must be boosted at least 140°F or higher on a daily basis.

Tempering Heated Water

The Plumbing Code requires that hot water be delivered at a temperature that avoids the likelihood of scalding. As water heated to 140°F or more can cause serious burns, water must be tempered before it is delivered to users though taps and other outlets.

The Plumbing Codealso sets maximum temperatures for water delivered to sanitary fixtures. For most types of building, the maximum temperature is 120°F for outlets such as basins, baths and showers (the limits are lower for buildings such as hospitals and schools).

Temperatures can be reduced to acceptable levels by installing:

- a tempering valve, or
- a thermostatic mixing valve.

A tempering valve is installed in the hot water line close to the cylinder and has a cold water connection to provide a pre-set hot water temperature at fixtures. Valves are factory pre-set but are able to be adjusted to cater for specific temperature requirements.

As water may be delivered at any temperature to non-personal hygiene fixtures such as sinks and laundries, a tempering valve is not required to these fixtures. However, if a wetback water heater or other uncontrolled heat source is used, tempering the supply to all fixtures is good practice.

Most dishwasher and washing machine manufacturers may require the installation of a tempering valve for warranty purposes where the unit does not heat its own water.

Specific Requirements For Gas Storage Water Heaters

Gas storage water heaters must have:

- adequate ventilation of the cylinder
- a flue to remove exhaust gases.

They must be:

- serviced annually
- flushed out regularly to remove water sediment at the bottom of the cylinder
- checked to ensure that vents are not blocked.

Controlling Pressure in Storage Cylinders

Most new hot water cylinders have reduced pressure from the water main system, but some buildings may still use low pressure systems.

Low Pressure, Open-Vented, Header Tank System

Low pressure, open-vented systems can be used to provide hot water supply for houses. In order to provide adequate hot water supply pressure, cold water is stored in a header tank located at a higher level than the storage water cylinder, from which water is gravity-fed into the bottom of the storage water cylinder.

As the water is heated, it rises to the top of the cylinder where it can be drawn off through taps or shower outlets and will be replaced from the header tank. The gravity feed provides the water pressure that pushes the water to the outlets, as long as they are at a lower level than the stored water.

As heating causes the water within the cylinder to expand, an open-vent pipe provides an outlet for excess pressure. The pipe usually feeds back into the header tank supply.

Low Pressure, Pressure-Reducing Valve System

Note: This system may be open-vented or unvented.

The low pressure, open-vented, pressure-reducing valve system works in the same way as the header tank system but uses a pressure-reducing valve to reduce the high pressure water from the main supply down to a pressure, or head, that is able to be maintained within the height of the vent pipe, which usually discharges above the roof.

This system, commonly known as an unequal pressure system, supplies low pressure hot water and high pressure cold water to fixtures. Its major disadvantage is that it is difficult to achieve balanced flow to a shower.

The unvented, low pressure system must also include a pressure relief valve.

District Main Pressure, Unvented Systems

District main pressure, unvented, storage water heater systems supply main pressure hot water to all outlets so both the hot and the cold water is delivered to outlets at the same pressure. An internal or external expansion vessel allows the heated water to expand, and systems must incorporate a pressure relief valve in case the expansion vessel fails.

All pressure cylinders require the pressure relief valve to be both pressure and temperature operated (temperature pressure relief or TPR) in order to provide a dual failure mode.

Valves must be specified for unvented systems to achieve the required pressure rating. As cold water storage header tanks and vent pipes are not required, the system allows greater flexibility in locating the cylinder.

Valves must be installed so they are:

- accessible for repairs and maintenance
- protected from damage
- protected from frost

Unqualified people should not alter, remove or dismantle valves on any potable water supply system.

Hot Water Pipes

Hot water pipes must be appropriate for the temperature and pressure of water being piped. Following are a number of factors that need to be considered:

Materials

Hot water pipes must be able to withstand the maximum temperature of the water being piped. Pipe material may be copper or an appropriate thermoplastic material.

Materials suitable for hot water supply pipes include:

- copper
- chlorinated polyvinylchloride (CPVC or PVCc)
- random polypropylene (PP-R)
- cross-linked polyethylene (PEX).

Toilets

In an average home, up to 30% of water use is for toilet flushing. This can be reduced by:

- ensuring a dual flush cistern is specified
- installing a water-efficient toilet pan
- using collected rainwater or treated greywater for flushing

- installing waterless composting toilets where no main sewer connection is available

Many older cisterns use 3 gallons of water, which is far more than is necessary. To reduce the amount of water used, replace the inefficient cistern with a modern dual-flush one. (A new pan may be needed where a dual flush cistern cannot be fitted to the existing one.)

If fitting a new pan/cistern is impractical, you can opt for the following to reduce water use:

- place an object such as a brick, or plastic milk bottle filled with water that has the top firmly screwed on into the cistern to reduce the amount of water required to fill an older cistern
- adjust the float ball by bending it down slightly to reduce the volume of water in the cistern – ensure that sufficient flow and volume is maintained for an adequate flush
- ensure that the cistern supply shuts off fully when not in use

In all cases, sufficient flow and volume must be maintained so the pan is cleared with a single flush.

Other Fixtures

Water usage can be reduced by specifying/installing:

- low-flow shower heads that use less than 2.5 gallons per minute and still deliver a comfortable shower
- aerators on taps used for hand washing

An aerator on a tap used for hand washing will reduce the flow while still providing plenty of water. Aerators should not be specified for taps on fixtures such as baths where a large volume of water is required.

Appliances

Reduce water use by recommending water-efficient appliances. Do not specify or install waste disposal units. Instead, where possible, encourage building users to compost all organic kitchen waste.

Wastewater

With wastewater, the overriding consideration is building users' health and safety.

Most buildings will connect to a town or city sewerage system. However, there are options for on-site disposal that can be used when there is no main sewer connection available. Some wastewater can also be recycled to reduce building water use.

In this section, we describe best practice options for installing standard wastewater systems. We also describe other options such as greywater recycling.

Building Design Considerations

There are several issues with sanitary plumbing and drainage that must be considered during building design.

Common Plumbing Coordination And Installation Issues

Accommodating water supply pipes in a building structure generally presents few problems, but this is not the case with drainage pipes. As they are generally larger and need to be installed with minimum gradients, they also take up more space.

Problems can occur because of a lack of coordination at the design stage, and/or between builder and plumbing contractor. Once the building is under construction, it is sometimes too late to achieve a proper solution. Be sure that these common plumbing coordination issues are resolved prior to the beginning of the plumbing installation:

- pipework needs to be surface mounted because there is insufficient space for it to be concealed
- minimum gradients are compromised because of insufficient depth in floor joists to accommodate pipe gradient
- dimensional requirements for floor waste gully traps are compromised because of inadequate floor depth
- structural members are compromised by oversized holes and notches because there is inadequate provision of ducts and bulkheads
- top and bottom plates and bracing are compromised to accommodate pipes in timber-framed walls
- fixtures are located with the waste outlet directly over a joist, bearer or beam
- pipe runs that are overlong and unnecessarily complicated in order to navigate non-penetrable building elements such as steel beams
- pipe and drain noises that cause disturbance to living and sleeping areas

Sanitary Plumbing Systems

This section discusses the compliance requirements for discharges from wastewater and soil fixtures.

Types of fixtures

Wastewater fixtures are all sanitary fixtures or appliances that receive wastewater and are not soil fixtures. Wastewater fixtures include hand basins, showers, baths, sinks and tubs.

Soil fixtures collect solid and liquid excreted human waste and include toilets, urinals, slop sinks and so on. Soil fixtures may discharge directly into a drain or to a discharge stack. Discharges and vents should comply with the size, material and performance requirements in your local Plumbing Code.

Water Traps with Single Discharge Outlet

Every fixture that discharges foul water must incorporate a water trap to prevent foul air from entering the building. The type of fixture determines the size of the discharge pipe and trap, and the size of the discharge pipe determines the minimum gradient required for the pipe, as defined in your local Plumbing Code.

Water traps must be:

- level and protected from freezing
- the same size as the trap arm

No S-traps, ball traps, drum traps, traps with moving parts, or traps with interior partitions are allowed.

The developed length of a fixture discharge pipe between the sanitary fixture outlet or a sanitary appliance (e.g. a washing machine) discharge and the water seal is dependent on the size of the pipe diameter. See your local Plumbing Code for requirements.

Water Traps with Multiple Discharge Outlets

A trap may serve multiple fixtures such as:

- two domestic sinks and one dishwasher machine (note that if one sink has a waste disposal unit installed, it must be trapped separately)
- two laundry tubs
- one laundry tub and a clothes washing machine
- two hand basins

Access points

Access points must be provided so that blockages can be cleared. These locations are defined in your local Plumbing Code.

Venting Discharge Pipes

All discharge pipes must be vented in accordance with your local Plumbing Code requirements.

- A vent pipe may be connected to a relief vent, a waste stack vent, or a branch vent.
- A waste stack must be vertical with no offsets.

Toilets – discharge pipes and venting

Toilets may or may not require venting, check your local codes for requirements. Toilets cannot discharge directly to a waste stack.

Air Admittance Valves

Air admittance valves (AAV) provide an alternative to running vent pipes to the outside of the building. When flowing water causes a reduction of air pressure within the system, the air admittance valve will open automatically, admitting air into the system. It will close again when the pressure in the system is equal to or greater than the external pressure.

Air admittance valves must be:

- used in accordance with manufacturer's instructions
- installed in an accessible, ventilated space where they are protected from damage, sun exposure and freezing
- fitted in an upright position

Floor Wastes

Dry floor wastes are provided to drain away accidental water spillage. They:

- may discharge directly outdoors
- must not be connected to a blackwater drainage system – water in the trap may be lost through evaporation admitting unpleasant odors
- must have a grating flush with the floor so they do not create a hazard

Pipe Materials

Acceptable materials for sanitary plumbing:

- are outlined in your local Plumbing Code
- must be 3rd party tested or certified
- must be marked by the manufacturer on all pipes and fittings

Drainage Systems

Compliance requirements for drainage features vary, and are outlined in your local Plumbing Code.

Traps

A plumbing trap is a curved pipe that is located underneath a plumbing fixture. A plumbing trap is an essential part of a building's plumbing for two main reasons; the first is that a trap prevents the entry of odors into a building, secondly, a trap also catches items that may cause major clogs or blockages if allowed to travel further into the drainage system. In most municipalities, the installation of plumbing traps on all plumbing fixtures is required.

Ventilation of drains

Drainage systems must be ventilated to transport sewer gas through the building to the open air space. Properly installed vents prevent siphoning traps and sewage gas leakage to the interior building. Ventilation requirements differ between codes. Be sure to check your local plumbing code to verify vent requirements for your region.

Pipe Sizing and Gradient

Pipe gradients are expressed as a percentage; 1% is 1 foot of fall over 100 feet of length. Subsequently, 2% would be 2ft over 100 feet of fall, and broken down further, this equals 1/4 inch over 1 foot of length. The minimum diameter for a drain is prescribed in your local Plumbing Code.

Calculating larger diameters, the size and gradient of a drain is based on the total of all discharge units that each section of the pipe carries. Each fixture type is given a rating derived from its expected discharge. Check with your local Plumbing Code to determine appropriate pipe sizes.

Drains must be laid:

- with the correct percentage of fall in the direction of flow
- so the diameter does not decrease in the direction of flow
- with a minimum diameter and gradient as set in accordance with local building standards
- so that they are supported on a firm bed along its entire length when laid underground
- so that no rocks or debris shall touch, support, or be in the first 12 inches of backfill over pipe

Materials For Drains

Materials and standards for drainage pipes are outlined in your local Plumbing Code.

Testing Of Drains

You should not cover the drains over until they have been inspected and tested for leaks by the local building department. The tests depend on the situation and the type of drain, but can include the following:

- water test
- air test
- gas test with smoke
- peppermint test

Access for Maintenance

Access point requirements differ between codes. However, access points must be provided:

- if a branch is greater than or equal to 5 feet
- at the end of all stacks
- at the beginning of the line
- at every change in gradient over 45°
- at every horizontal change in direction more than 45°
- at every junction that serves a soil fixture or any branch drain longer than 2 m
- at least every 100 feet on straight drains
- at the junction of the building drain and building sewer (brought to finish grade level)

On-Site Wastewater Treatment

On-site wastewater treatment is an option if there is no sewer available or if the owner wants to recycle water to reduce demand on main wastewater systems. Some municipalities require that if a sewer is available the drainage system must be connected to it – although the local authority can provide a waiver, allowing building owners to reduce demand on main systems by recycling greywater or using a composting toilet. If there is no sewer available, on-site treatment is the only option available. With all wastewater disposal or recycling, health and safety must be the overriding priority.

Blackwater and Greywater

Blackwater is wastewater from toilets, dishwashing machines and sinks – because the fats, detergents and cleaning agents used in kitchen wastewater, this is considered blackwater and must be discharged accordingly. Greywater is waste from baths, showers and hand basins. Wastewater from clothes washing machines can be considered in either category.

On-Site Wastewater Treatment Options

Most on-site wastewater treatment systems involve two stages of treatment – the first stage in a tank or treatment system, and the second when the effluent is dispersed on to land or the garden and further breaks down. The first stage may be carried out in a septic tank or in a more advanced system such as an aerated wastewater treatment system or advanced sewage treatment system. These systems are much more advanced than septic tanks and treat effluent to a level that allows it to be used on the garden or even recycled for toilet flushing and vehicle washing.

In many areas, septic tanks are no longer permitted as there is not enough land available to safely treat the effluent after it is discharged from the tank. You can alternatively use treated greywater for irrigation or flushing toilets or use a composting toilet that is used in combination with a greywater treatment system. Both of these options reduce the amount of effluent to be disposed of by the wastewater treatment system.

Designing an On-Site Wastewater Treatment System

The appropriate wastewater treatment system will depend on the site, the required capacity, and compliance requirements.

Site Considerations

Site features that must be considered when designing a wastewater disposal system include:

- Is the water table high or low?
- Does the land become saturated during periods of high rainfall?
- What area of land is available for the system?
- What is the soil's ability to absorb moisture, e.g. is the subsoil clay, sand, loam etc?

Note: The amount of land needed for the disposal field depends on the soil.

Local municipalities may have their own regulations controlling the installation of wastewater treatment systems. Contact your local authority for approval before you begin to design a system.

Greywater and blackwater may be treated separately. If separate treatment systems are selected, you should include a means of diverting the greywater to the blackwater treatment system. A composting toilet eliminates the need for a blackwater system but must incorporate a means of dealing with urine. The greywater can be dispersed to a land application or used for irrigation.

System Capacity

Allow for a daily output of 50 gallons per person when designing a system. This will enable it to cope with peak discharge rates or temporary overloads. It must also be able to retain the total flow for at least of 24 hours.

The system should also have enough capacity for 3–5 years of sludge at the following rates:

- for blackwater and greywater – 20 gallons per person per year

- for blackwater only (where there is separate greywater system) – 15 gallons per person per year
- for greywater only – 10 gallons per person per year

The septic tank system capacity must allow for:

- variations in the quantity of effluent to be disposed of
- the possibility that householders will not manage and maintain the system effectively

Design Checklist

Use this guide to help assist you in your design of an on-site wastewater treatment system.

Stage 1: Feasibility Study

Determine requirements by preliminary:

- discussion with system suppliers
- identification of site conditions
- discussion with regulatory authorities

Stage 2: Evaluation And Investigation

Includes soil evaluation for permeability and drainage capability, and site investigation to identify:

- gradients
- site boundaries
- building positions
- runoff
- stability
- groundwater conditions
- location of nearby wells, springs and watercourses

- vegetation
- climatic conditions.

Stage 3: Selection System

Consider the:

- number of occupants
- site conditions
- local authority regulatory requirements.

Stage 4: Design Distribution System

Must include:

- uniform, effective and continuous distribution
- an area that can be controlled and maintained.

Stage 5: Approvals

Obtain as required from regulatory authorities for:

- location
- system specification
- operation
- maintenance
- monitoring

Stage 6: Installation

In accordance with:

- approved design drawings and specification
- manufacturer's instructions
- certification as required by the regulatory authority

Stage 7: Operation And Maintenance

Includes:

- providing an operations and maintenance manual
- keeping records of all servicing/inspections/remedial actions
- carrying out regular inspections/cleaning
- having a service and maintenance manual with the supplier (for AWTS/ASTS)
- maintaining the land application area.

Septic Tanks

A septic tank is a primary treatment system – that is, treatment of wastewater is minimal and involves only separation of solids and some preliminary anaerobic (without oxygen) action. Septic tanks provide minimal treatment for wastewater and are no longer allowed in many areas.

How Septic Tanks Work

Wastewater flows into the septic tank where solids and liquids separate. Partially decomposed solids settle to form sludge on the tank floor, and lighter materials such as fat and grease form a floating layer of scum. Effluent, which may still contain small particles of solids, flows out of the septic tank to a land-application disposal area. It filters through the soil, where it is treated by bacterial action.

The partially decomposed solids that settle on the bottom of the tank must be pumped out approximately every 3–5 years, depending on use. The septic tank outlet should be below the level of the floating scum layer so the amount of these solids that are dispersed onto the land is limited.

Septic tanks are generally gravity-fed. They must therefore be installed below the level of the house. If this is not possible, waste must be pumped to the tank. Tanks may incorporate tees or baffles at the inlet and outlet pipes to slow incoming wastewater and reduce sludge disturbance. Gas baffles may be incorporated to deflect gas from escaping through the outlet.

Construction and installation

Septic tanks can be factory-built and manufactured from reinforced cement mortar, fiberglass, steel, or plastics such as polyethylene and polypropylene. Tanks can be installed underground, or above ground.

Below-ground systems are the most common method for septic tanks. These systems include:

- resist loads from the surrounding soil and groundwater
- resist hydrostatic uplift (tendency to float)
- prevent surface and groundwater getting in
- have inspection covers that are not accessible to children
- be clear of trafficked areas.

An above-ground system may be required if the site is sub-ground rock or has a very high water table. Above ground systems must:

- be watertight
- be durable
- be UV resistant
- resist earthquake forces
- have inspection covers that are not accessible to children

Aerated and Advanced Wastewater Treatment Systems

- These systems treat sewage to a higher level than septic tanks, allowing it to be safely used for gardening or even recycled for toilet flushing.

- Aerated water treatment systems (AWTS) and advanced sewage treatment systems (ASTS) are secondary treatment systems, that is, they involve both anaerobic and aerobic (with oxygen) treatment to a higher level than a primary treatment system, resulting in effluent that is suitable for garden (excluding fruit and vegetables) and landscape irrigation.
- At the highest level of treatment (from ASTS), the treated effluent can be used in non-potable situations such as toilet flushing, vehicle washing and firefighting.
- In many areas, a secondary treatment system is the only option permitted.

How an AWTS Works

Water flows through a series of chambers or tanks that progressively treat and filter the wastewater. The first chamber is an anaerobic treatment chamber where settlement of solids and anaerobic decomposition occur. Effluent passes through filters into chamber 2. This is where aeration treatment occurs and is pumped in air and creates turbulence and aerates effluent.

The chamber may incorporate a bioreactor to give additional bacterial treatment. Next, the effluent passes through a second, finer filter to the clarification chamber. In this chamber, fine sludge particle settle and are pumped back to the first chamber. In chamber 4, a submersible pump distributes treated effluent to the disposal field.

After treating wastewater in a similar process, an advanced treatment system may pass the effluent through a sand filter, a packed bed filter or a textile bed reactor, where effluent trickles through the bed material containing micro-organisms that treat any remaining fine solids before being pumped to the disposal field.

Land-Application Disposal System

Following treatment in a septic tank or other treatment system, effluent is disposed of on land. There are several ways this can be done.

Following primary or secondary treatment, effluent is moved by gravity or pump via a subsoil drainage system to a land-application disposal area, where bacterial action carries out the final treatment as the effluent filters through the soil. The effluent from an Advanced Sewage Treatment System will most likely have significant treatment and can be used for irrigation. It is not however suitable to be used for crops for human consumption. The various land-application disposal systems (listed above) each have their own advantages and disadvantages.

Gravity Soakage Trenches/Beds (Septic Tank System Only)

Perforated dose lines, are laid in trenches or beds filled with aggregate and covered with a layer of topsoil. Effluent trickles through the aggregate into the surrounding soil.

Gravity soakage can only be used with a septic tank system. It can work well in reasonably flat, good draining soils, but a common problem is that the effluent flow does not spread evenly over the disposal area, and most of the effluent will discharge at the beginning of the trench. If a trench is too deep, aerobic bacterial treatment of effluent will not occur.

Low Pressure Effluent Distribution (LPED)/Dose Loading (Septic Tank System Only)

Effluent may be discharged more evenly across the disposal area by pump or dose loading. A controlled dose is pumped through a doseline at regular intervals over a 24-hour period, ensuring the effluent is spread over the whole area and also gets a rest period between soakings. It also eliminates the chance of disposal surges that may occur during periods of high household use.

Alternatively, even soakage can also be achieved by nesting the perforated dose line within a drainage coil installed in the trench. The effluent moves along the drainage coil, spreading more evenly across the whole land-application area.

A distribution or diverter box can be used to different parts of the field and allow the trenches to be periodically 'rested' to prevent drains becoming clogged by saturated conditions. As the soil filtering process provides the secondary treatment, the disposal area for both gravity soakage and LPED disposal should be fenced off to prohibit access.

Drip-Line Irrigation Systems

Several options of drip-line irrigation are available including:

- **Sub-Surface Drip** – where lines are buried in topsoil
- **Surface Drip** – where lines are laid on the surface and covered in bark or mulch
- **Spray System** – where treated and disinfected effluent is sprayed over the ground surface.

Drip-line systems are only suitable for secondary treatment effluent. Effluent is pumped, distributing it over the whole of the effluent field each time it operates.

Evapo-Transpiration Systems (ETS)

In evapo-transpiration, effluent is dispersed into beds planted with selected, shallow-rooted plants. The plants absorb effluent through the roots and release water through the leaves into the atmosphere in a natural process of transpiration. Effluent not taken up by plants will be absorbed into the soil.

Sand mound systems

Where it is not possible to achieve a suitable depth of trench due to a high natural water table or poor percolation, soil or sand can be mounded to provide a suitable filtering depth for the effluent treatment. The effluent can trickle through the mound into the underlying soil.

Treated effluent from an on-site domestic waste-water system may be discharged through an evapo-transpiration trench. The effluent is absorbed into the topsoil, taken up by plants, or may evaporate. Traditional distribution of effluent from a septic tank has been through gravity-fed perforated pipes in an aggregate bed.

Site Investigation

Site requirements for an effluent disposal system must consider the:

- nature of the subsoil, including permeability (the rate at which water can percolate through it) and stability
- characteristics of the site such as:
    - slope
    - natural drainage characteristics
    - water table levels
    - water course location
    - tendency to flood
    - area available for land application
    - vegetation and planting
- potential effects on:
    - downstream neighboring properties
    - natural water courses the sea

- local ecology
- the field location i.e. fields must not be grazed or driven over.

Maintenance and Problems

Owners are legally responsible for maintaining their on-site wastewater treatment system.

Maintaining Septic Tank Systems

- Inspect tank annually for sludge and scum levels.
- The tank should be pumped out approximately every 3–5 years. Have tank pumped out when:
  - the top of the floating scum is 3 inches or less from the bottom of the outlet
  - sludge has built up to within 10 inches of the bottom of the outlet
- Check and clean outlet filters regularly (6-monthly)
- Alternate dispersal to the land-application areas approximately every 3–6 months.

Maintaining Aerated Water/Advanced Sewage Treatment System

AWTS and ASTS should be serviced by a qualified service technician, generally every 6 months, to:

- clean or replace filters as required
- monitor the effluent quality, including pH level, of the first chamber
- check the submersible pump and float switch operation
- record all inspection maintenance and monitoring events
- replace the submersible pump at 7–10 yearly intervals.

Maintaining Land-Application/Disposal Systems

- Keep the area clear of deep rooting trees and shrubs (these may grow into and cause blockage of the system).
- Clean and service pumps, siphons and filters according to manufacturers' instructions.
- Flush drip lines regularly to remove accumulated sediment.
- Redirect effluent periodically to alternative trenches or beds (septic tanks).
- Mow grass and maintain plants in evapo-transpiration areas.
- Ensure that surface water drains around land-application areas are kept clear to reduce rainwater runoff into trenches or beds.

Avoiding Problems

- Specify water-efficient appliances.
- Prevent overloading the system by minimizing water use (e.g. spread heavy water-use activities such as clothes washing over several days) and installing a separate greywater treatment system.
- Strong chemicals restrict the biological action within the tank – select cleaners and washing products that do not hamper the decomposition process, and make sure chemical products such as volatile thinners, bleaches and disinfectants do not enter the system
- Kitchen waste should not enter on-site wastewater treatment systems – compost kitchen waste instead of installing a garbage disposal unit.
- Systems cannot deal with condoms, dental floss, tampons, sanitary napkins, etc. – these should be wrapped up and discarded in the waste disposal service.

Signs of Trouble

Following are a few signals that the system is not working properly.

- If there is a foul smell around tank or land application area
- If the tank overflows
- If the ground around the tank is soggy
- If sinks/basins/toilets are emptying slowly
- If fixtures make a gurgling noise when emptying
- If the grass is unusually dark green over the land application area
- If black liquid is oozing from the trenches

Solving problems

- If the tank is too full, have it pumped out.
- If the tank contains too much sludge and scum, have it pumped out, or desludged.
- If there is too much water going into the tank, use less water and check for stormwater infiltration.
- If toxic chemicals are going into the system, reduce use of hard detergents/cleaners.

Greywater Recycling

Greywater is wastewater from bathrooms, diverted for garden irrigation or (if the greywater is treated) for toilet flushing.

Greywater is water from sink basins, baths and showers. It can be recycled for use in garden irrigation and, if treated, for toilet flushing. Recycling greywater can help reduce the load on sewerage systems including on-site treatment systems and provide a garden water supply, reducing demand on other sources of water.

Safety Considerations

Greywater recycling must be designed and installed with care as it is potentially unsafe in some situations. To reduce risk, greywater used for garden irrigation should deliver water below the soil surface. Be aware that greywater recycling carries potential health risks and should not be used in homes.

It is not recommend to use greywater for:

- washing clothes
- garden irrigation by sprinkler
- use on vegetables or salad plants

Collected rainwater is a more preferred option for toilet flushing than greywater. Greywater systems must be designed so that any overflow can be discharged to a sewer or on-site blackwater treatment system. Laundry water may also be classified as greywater but, as it may contain detergents and other contaminants, it is not recommended for recycling.

Greywater Systems

A greywater system diverts waste water either to irrigation or a treatment and recycling system.

The key consideration for installing a waste water system is the health and safety of property users. Greywater systems used for irrigation typically comprise a surge tank and a method of discharge to an irrigation system. Greywater systems used for toilet flushing should have a treatment system. Greywater systems must comply with local Code requirements.

How a Greywater System Works

Greywater is water from sink basins, baths and showers that is piped to a surge tank. The greywater is held briefly in the tank before being discharged to an irrigation or treatment system. The greywater can be diverted either by gravity or by using a pump.

The surge tank can be any type of container that is suitable for holding (but not storing) the initial surge of water. The surge tank must be emptied completely each time greywater is dispersed to the irrigation or treatment system – greywater must not sit for extended periods of time in the tank.

A gravity system can only be used when there is sufficient fall from the laundry/bathroom drain to the surge tank.

The surge tank should:

- be vented
- have a trapped overflow
- discharge directly into the sewer or to an on-site discharge
- be sealed
- be vermin proof.

A three-way valve manually diverts water from the normal drainage system to the surge tank. The machine discharge pipe must not be more than 12 inches above the top of the machine to avoid overloading the pump, and it must discharge into a 1 -1/2 inches open pipe to avoid the possibility of water being siphoned from the machine. This can only be done if there is sufficient distance between the floor level and the outside ground level to allow a gravity feed to the surge tank. It entails a valve on the appropriate waste pipes to divert the wastewater to the surge tank. The pipes may be individual or combined wastes from the laundry and bathroom (but not from the kitchen). Water is then pumped to the irrigated area.

A pumped system, using a simple submersible pump and float switch, must be used where there is insufficient fall. If necessary, the surge tank may be partially or wholly below ground level.

Treatment Of Greywater

Treatment of greywater may include:

- filtering
- settlement of solids
- flotation and separation of lighter solids
- anaerobic or aerobic digestion
- chemical or UV disinfection.

Greywater used for irrigation should be filtered as it still contains high levels of solids and is otherwise likely to clog the irrigation system.

Filtering may be:

- a filter to catch the lint, or
- use of large diameter pipes that allow solids to pass through the system without causing blockages

Greywater must be filtered to avoid clogging the system. In a simple filtration device such as this, greywater is discharged into a tank containing the filter material that consists of a layer of pine bark over a filter-cloth and a sand layer. The water flows continuously through the filter and directly to the irrigation system. Greywater filters will need to be replaced from time to time, and the solids that settle on top of the greywater must be removed regularly. Greywater should only be used for toilet flushing if it has been treated to reduce harmful bacteria to an acceptable level.

Designing A Greywater System

When designing a greywater system you should consider the:

- personal habits of the users i.e. what they put in the system
- quantity of wastewater output
- size of the site

- soil conditions of the site
- type of recycling usage required i.e. whether it is for irrigation only, or for re-use within the home

Note: A greywater recycling system does not allow a reduction on on-site treatment capacity as greywater may still enter the treatment system when the storage tanks are overloaded.

Installations that are designed in accordance with local greywater codes are suited to a greywater installation because:

- greywater and blackwater systems are separate until they are outside the building
- greywater intended for recycling can be directed to a single gully trap where it can easily be diverted for re-use
- other wastewater, such as kitchen wastewater, can be directed to a separate gully

Proprietary Greywater Systems

Commercially manufactured systems that treat greywater to a standard for toilet flushing and/or irrigation are also available.

Commercially manufactured units suitable for irrigation typically comprise of:

- a submersible pump that automatically pumps the greywater to the irrigation system
- either a manual or remote electrically operated over-ride switch that diverts all the greywater to the sewer if necessary
- a partially self-cleaning filter.

Because greywater used for toilet flushing should be treated to reduce harmful bacteria, many commercially manufactured systems that do this are available.

Irrigating With Greywater

When using greywater for irrigation, it's important to comply with local authority requirements and to ensure that the greywater is used safely.

Local Authority Restrictions

Check with the local authority for restrictions on greywater use such as:

- minimum distance of discharge from boundaries, waterways, wells and bores, and sea
- maximum allowed daily discharge rate

Safety and Health

Greywater that has not been disinfected will contain bacteria. Greywater may also contain chemicals from cleaning products, detergents and bleaches that can contaminate the soil and kill plants. If using greywater for irrigation, avoid:

- harsh detergents, softeners and whiteners
- bleach or cleaners with chlorine
- cleaners containing boron

Minimize any health risks:

- Do not allow greywater to pond.
- Do not spray, as this creates aerosol droplets that can drift.
- Discharge greywater below the soil surface.
- Do not use for vegetables – it is suitable for shrubs, flowers and fruit trees only.
- Do not irrigate near children's play equipment or play areas.
- Do not irrigate plants that prefer acidic conditions.

Distribution Systems

A greywater distribution system should incorporate a distribution box and branched drain network, so that water can be diverted to different parts of the dispersal area, to allow each area to rest.

These are filled with large aggregate metal and rely on absorption and/or transpiration, or mulch in areas of good soil permeability.

Mulch-filled swales are channels filled with pine bark mulch. Water is discharged through small diameter slotted pipes. The mulch prevents physical contact with the water. There are several manufactured systems available, typically using a pump to distribute water.

Composting Toilets

Composting toilets can be used where there is no main sewer connection or in some circumstances to reduce demand on the sewer systems.

How Composting Toilets Work

Flushing toilets account for approximately 30% of domestic water use. Composting (or waterless) toilets eliminate the need for flushing water.

A composting toilet breaks down human waste and other added organic material by an aerobic process in the same way that garden compost is made. The end product should be odorless, soil-like humus that can be buried on-site.

For the composting to occur, the moisture content in a composting toilet must be minimal. This generally requires separating out the urine by evaporation or a separate collection system.

The urine can then be disposed of by:

- a septic tank or other on-site blackwater treatment system
- a rock filled soakaway/soakpit
- a storage tank – it can be used as a fertilizer for citrus trees

Approval from your local municipality for the proposed disposal method must be obtained prior to beginning its construction.

Requirements of Composting Toilet System

Composting toilets must contain and act on pathogens. They must:

- meet good sanitation and public health requirements
- keep human and pet contact with effluent to a minimum
- prevent contact with disease carriers, such as flies
- produce no offensive smells
- end with an inoffensive product with low concentrations of harmful bacteria.

In addition, they:

- can only be installed where there is no available public sewer connection
- require continuous extract ventilation
- may require a separate urine disposal system
- must have an alternative means of disposal for maintenance or repair of the system

Deciding To Install a Composting Toilet

They require a significant commitment to monitor and maintain the system, which involves:

- changing the full bins with batch systems
- removing the compost at regular intervals from continuous systems
- adding soil, burying or disposing of the compost
- cleaning the system as required by the manufacturer.

If a composting toilet is not properly maintained and monitored, the end product may not be properly composted, which means:

- removal and cleaning may be unpleasant
- there may be a health risk
- there may be odors.

Types of Composting Toilet

A range of manufactured composting toilets are available – these have the benefit of research and development input and have been tested over many years. Composting toilets may be batch type or continuous.

Batch-Type Units

Most small self-contained composting toilets are of the batch type. These have two or more bins. When one bin is full, it must be moved to a suitable place (generally outside) to allow composting to be completed. This generally takes 5–6 weeks. The material is buried, an empty bin replaces the full one and the process is repeated.

They generally do not have the underfloor space requirement of continuous systems, which makes planning easier particularly if more than one toilet is required.

Batch type systems must be vented and may:

- be heated
- have mechanical ventilation that requires an electrical connection – 12-volt models are available and power consumption is generally low
- require a vent pipe
- require a connection to an on-site blackwater outfall for separated urine
- require the toilet room to be maintained at a regular temperature.

Continuous Systems

Continuous systems have one chamber where all waste is received and stored until composting is completed. The finished compost is removed and buried.

A continuous system requires an underfloor space, and each pan must have its own chamber. (Therefore, having more than one toilet may be difficult and expensive.)

They may require:

- a positive air pressure in the toilet room to avoid smells
- an air inlet and exhaust which may be driven by convection, electric fan or solar heat (generally power consumption is low)
- a means of draining excess liquid
- access to a hatch for removal of finished compost
- the addition (by the users) of organic bulking agents such as sawdust to aid the decomposition process.

Stormwater Control and Landscaping

Careful stormwater control and landscaping can cut water use, reduce demand on main stormwater systems, and protect waterways from contamination.

Stormwater must be managed to minimize the risk of flooding. But allowing all stormwater from a property to run into drains not only wastes a potential source of garden water but also means that contaminants such as oil, paint and animal droppings are carried into waterways.

Stormwater can instead be harnessed for irrigation or otherwise disposed of on-site, reducing demand on property's water supply. With an estimated 10-30% of household water used for gardening, this can have benefits in terms of water efficiency.

Similarly, garden water use can be reduced by selecting plants that require little water, by collecting and using rainwater (which also has the effect of reducing stormwater runoff), and/or by using greywater for garden irrigation.

Controlling Stormwater Runoff

Stormwater may be disposed of:

- into a natural watercourse
- into a water storage tank
- into a soakpit

In a natural landscape, rainwater surface runoff averages vary. However, in urban areas, due to the increase in hard surfaces such as roads, driveways and reduced vegetation, average runoff is fairly high. Runoff can cause pollution by carrying soil, contaminations (such as fuel) from roads and vehicles, human and animal waste, and chemicals (e.g. fertilizers, pesticides, industrial chemicals and household cleaners) into waterways.

Stormwater runoff can be reduced by:

- collecting and storing rainwater in storage tanks
- using permeable paving
- incorporating swales to slow the rate of surface water movement
- installing a green roof as part of a new building design

Rainwater Storage

Stormwater runoff can be reduced by collecting and storing rainwater for gardening, toilet flushing or other uses. This also has other benefits, such as reducing water costs for properties on metered supply, and reducing demand on other stormwater disposal systems.

Permeable Paving

Stormwater runoff rate can be reduced by using permeable paving for driveways, footpaths and parking areas instead of hard, impervious paving such as asphalt or concrete.

Permeable or porous surfaces include:

- gravel
- concrete-grass paving
- porous concrete/asphalt
- open-jointed paving over gravel
- green roofs.

Permeable surfaces are best suited to:

- areas with a slope of less than 6%
- low traffic volumes
- low speed traffic.

Stormwater runoff will be slowed by open paving blocks that allow water to infiltrate through gaps. Grass is usually planted in the gaps.

Paving blocks on a sand base and with open joints can be used to slow rainwater runoff but they are not as permeable as concrete/grass paving options.

Swales

Swales are wide, shallow drainage channels running across the slope of the ground that forms part of the landscape. They reduce runoff rate by:

- retarding the flow rate of surface water
- providing a means of infiltration into the subsoil

Swales should be wide and shallow, with a gradient across the slope of at least 2%. On steeper slopes, they should include a check dam to slow the flow rate.

Run-off is slowed and absorption of rainwater into the ground is increased by use of a check dam on sloping ground.

Green Roofs

A green roof has vegetation planted into a layer of growing/drainage medium lain over a waterproof membrane. Green roofs help to reduce the water runoff rate by retaining the water, which is then lost through slow drainage, transpiration and evaporation.

Pitched roofs suitable for use as green roofs, but lower pitches require less depth of growing medium. Steeper roofs require a deeper layer and measures to prevent erosion. Obtain specialist advice before deciding on the type of growing medium and plants to use.

Disadvantages of green roofs include:

- they cost more
- the roof is heavier than a conventional roof and will require additional structure for support
- they require specific design under the Building Code
- the membrane, flashings and rainwater collection and retention system must be carefully designed and detailed to prevent deterioration from plant roots and damp conditions
- if the roof leaks, it may be difficult to find the cause, and the cost of repairing it will be expensive
- they are likely to require more maintenance than a conventional roof
- in arid regions, they require an irrigation system for the plants, which will increase the cost.

Advantages of green roofs include:

- protection of the waterproof membrane by the growing medium from ultraviolet light
- they reduce the rate of rainwater runoff
- good noise reduction
- a natural appearance that blends into the landscape.

Reducing Garden Water Use

The amount of water needed for garden irrigation can be reduced significantly by selecting plants that require little water. Opt for plants that grow well with limited or no watering within the local area.

Garden water use can also be reduced by using an efficient irrigation system. You should look for one that applies a smaller volume of water directly where it is needed, such as by drip irrigation rather than sprinklers. Water use should be controlled by the use of timers (for instance, to water in the evening to reduce the amount of water evaporation) and/or in-ground moisture meters.

Other ways to reduce garden water use include storing and using rainwater for irrigation, using greywater for irrigation, and diverting stormwater for irrigation instead of allowing it to run into drains.

Conclusion

Water consumption is the need of the hour. You need to make the most of the water available to you and use it efficiently, not only to reduce costs, but also to preserve it for the future. Better systems like more efficient flushing systems and water recycling systems have made it relatively easier to conserve water. Opt for solutions that work best in your specific situation and comply by the standard and local building codes.

BIBLIOGRAPHY AND REFERENCES

American Concrete Institute. (2011). *ACI 318-Building Code Requirements for Structural Concrete.* Farmington Hills, MI.

American Concrete Institute. (2011). *ACI 530- Building Code Requirements for Masonry Structures.* Farmington Hills, MI.

American Forest and Paper Association. (2011) *Wood Frame Construction Manual for One- and Two-Family Dwellings (WFCM).* Leesburg, VA.

American Iron and Steel Institute. (2007). *Standard for Cold-Formed Steel Framing—Prescriptive Method for One - and Two-Family Dwellings (AISI S230).* Washington D.C.

American Society of Heating, Refrigerating and Air-Conditioning Engineers. (2006-2009). *ASHRAE Handbook: Fundamentals.* Atlanta, GA.

Ballast, D. (1994). *Handbook of Construction Tolerances.* McGraw Hill.

Building Industry Association of San Diego County. (1993). *Top 25 Construction Problems and Their Resolution.* Construction Quality Task Force.

California Building Industry Association. (2005). *SB 800, The Homebuilder "FIX IT" Construction Dispute Resolution Law.* Sacramento, CA.

California, State of, Department of Real Estate. (1996). *Operating Cost Manual for Homeowner Association.* Sacramento, CA.

California, State of, Contractor's State License Board. (1982). *Workmanship Guidelines.* Sacramento, CA.

Concrete Committee of San Diego County. (2001). *Concrete Performance Standards and Maintenance guidelines.* San Diego, CA.

Gypsum Association. (2012). *Fire Resistance Design Manual.* Hyattsville, MD.

Hansen, D. & Kardon, R. (2011). *Code Check – Building.* Taunton Press. Newtown, CT.

Hansen, D. & Kardon, R. (2010). *Code Check – Electrical.* Taunton Press. Newtown, CT.

Hansen, D. & Kardon, R. (2011). *Code Check – Plumbing &Mechanical.* Taunton Press. Newtown, CT.

International Code Council. (2007). *California Building Code.* Whittier, CA.

International Code Council. (2007). *California Electrical Code.* Whittier, CA.

International Code Council. (2007). *California Mechanical Code.* Whittier, CA.

International Code Council. (2007). *California Plumbing Code.* Whittier, CA.

International Code Council. (2006-2009). *International Residential Code for One and Two Family Dwellings.* Washington D.C.

International Association of Plumbing & Mechanical Officials. (2009). *Uniform Mechanical Code.* Ontario, CA.

International Association of Plumbing & Mechanical Officials. (2009). *Uniform Plumbing Code.* Ontario, CA.

Journal of Light Construction. (1997). *Troubleshooting Guide to Residential Construction*, Builderburg Group.

NAHM Research Center, Inc. (2001). *Mold in Residential Buildings.* Washington D.C.

National Association of State Contracting Licensing Agencies. (2009). *NASCLA Residential Construction Standards.* Phoenix, AZ.

National Fire Protection Association. (2011). *National Electrical Code.*

National Roofing Contractor's Association. (2007-1009). *NCRA Roofing and Waterproofing Manual.* Vols 1, 2, & 3. Rosemont, IL.

National Wood Flooring Association. (2000). *Problems, Causes and Cures.* Ellisville, MO.

NAHB Home Builder Press. (2005). *Residential Construction Performance Guidelines.* Washington D.C.

New Jersey, State of, Division of Codes and Standards. (2005). *Homeowners booklet,* New Home Warranty Program. NJ.

Reynolds, D. (1998). *Residential & Light Commercial Construction Standards.* R.S. Means, Inc. Kingston, MA.

Sacks, A. (1994). *Residential Water Problems.* NAHM Home Builder Press. Washington, DC.

Structural Building Component Association & Truss Plate Institute. (2006-2013). *Guide to Good Practice for Handling, Installing, Restraining & Bracing of Metal Plate Connected Wood Trusses.*

Tenebaum, D. (1996). *The Complete Idiot's Guide to Trouble Free Home Repair.* Alpha Books. NY.

Truss Plate Institute. (2008). *National Design Standard for Metal Plate Connected Wood Truss Construction.* Alexandria, VA.

ABOUT THE AUTHOR

Ryan Brautovich is an Army veteran with more than 20 years of home construction, home remodeling and building experience who has consulted for Fortune 500 home builders as well as the Top 100 privately held home building companies. He is a custom home builder in California and a California licensed general contractor. Ryan is International Code Council Certified, an International and California Building Inspector as well as an International and California Plumbing Inspector. He is a graduate of Auburn University with degrees in both Accounting and Business Management. He has consulted for the City of Lancaster (CA) Building & Safety Department, K. Hovnanian Homes, Beezer Homes, Pardee Homes, KB Homes, Standard Pacific Homes, American Premiere Homes, Richmond American Homes, DR Horton, and Frontier Homes – just to name a few.

Ryan founded the Construction H.E.L.P. Foundation, a national nonprofit organization, dedicated to advocating for and meeting the needs of individuals who have suffered at the hands of unscrupulous contractors and sub-contractors who simply took advantage of the helpless homeowner in order to make a quick buck – and either didn't finish the project, overcharged or simply took money and didn't perform the work as promised. Over the years, the number of phone calls Ryan received increased dramatically from frustrated and angry homeowners who were desperately seeking help after being ripped off by other contractors. As a result, he founded the Construction H.E.L.P. Foundation, and it's educational assistance program – Home Construction Audit – to provide assistance and education to homeowners. As the founder of the Construction H.E.L.P. Foundation, Ryan has made it the organization's daily mission to return ethics to the home building and home remodeling profession and provide homeowners with the expert help and crucial knowledge they need so that they will never be taken advantage of again.

www.ingramcontent.com/pod-product-compliance
Lightning Source LLC
Chambersburg PA
CBHW030241170426
43202CB00007B/87